Dear Math-Minded Families,

Thank you and congratulations for taking your first step towards investing in math mastery for your child(ren). For many, abacus-based math instruction is a new way of doing math. With that in mind, I have designed a very fun and user-friendly system for young children. Be sure that you have your jr.counter (www.mathjr.org) and consider joining our Abacus Math Club (www.mathjr.org).

The Mathematician, Jr. system includes:

- An introduction to the abacus using a simplified abacus called the jr.counter (by sure to purchase yours at www.mathjr.org)
- Ample practice for each skill
- Mental Practice
- An emphasis on positive message using adinkra symbols

You will find that using the jr.counter for calculations, and mental practice will significantly develop and improve not only your children's math ability, but will also enhance their understanding of the fundamentals of math language and number relationships. Additionally, your child will be able to conduct mental calculations and will be able to give a better explanation to how he/she has arrived at their answer.

To help your child's commit to increasing math learning, it would be a great idea to add incentives to the contract on the next page and have all 'parties' sign. This contract will help keep your child motivated.

With Much Math Success,

You Abacus Math Club Guide, Dr. Ameerah Anakaona
Founder of Mathematician, Jr.

The Mathematician's Contract

I, _____ (Mathematician, Jr.) promise to practice each day so I can become a super smart mathematician. I will work in a place where I can focus. I will also be sure to focus very hard when it comes to mental math practice math.

_____   _____
Mathematician, Jr.   Date

_____   _____
Parent/ Witness   Date

Mathematician's Parent Contract

After _____ (Mathematician's name) successfully completes Math Readiness Workbook Level I, I _____ (Parent/ guardian) promise to (write the reward that your child will receive for completing this book):

_____
_____
_____.

_____   _____
Future Mathematician, Jr.   Date

_____   _____
Parent/ Witness   Date

AFRICA IS A BEAUTIFUL PLACE!

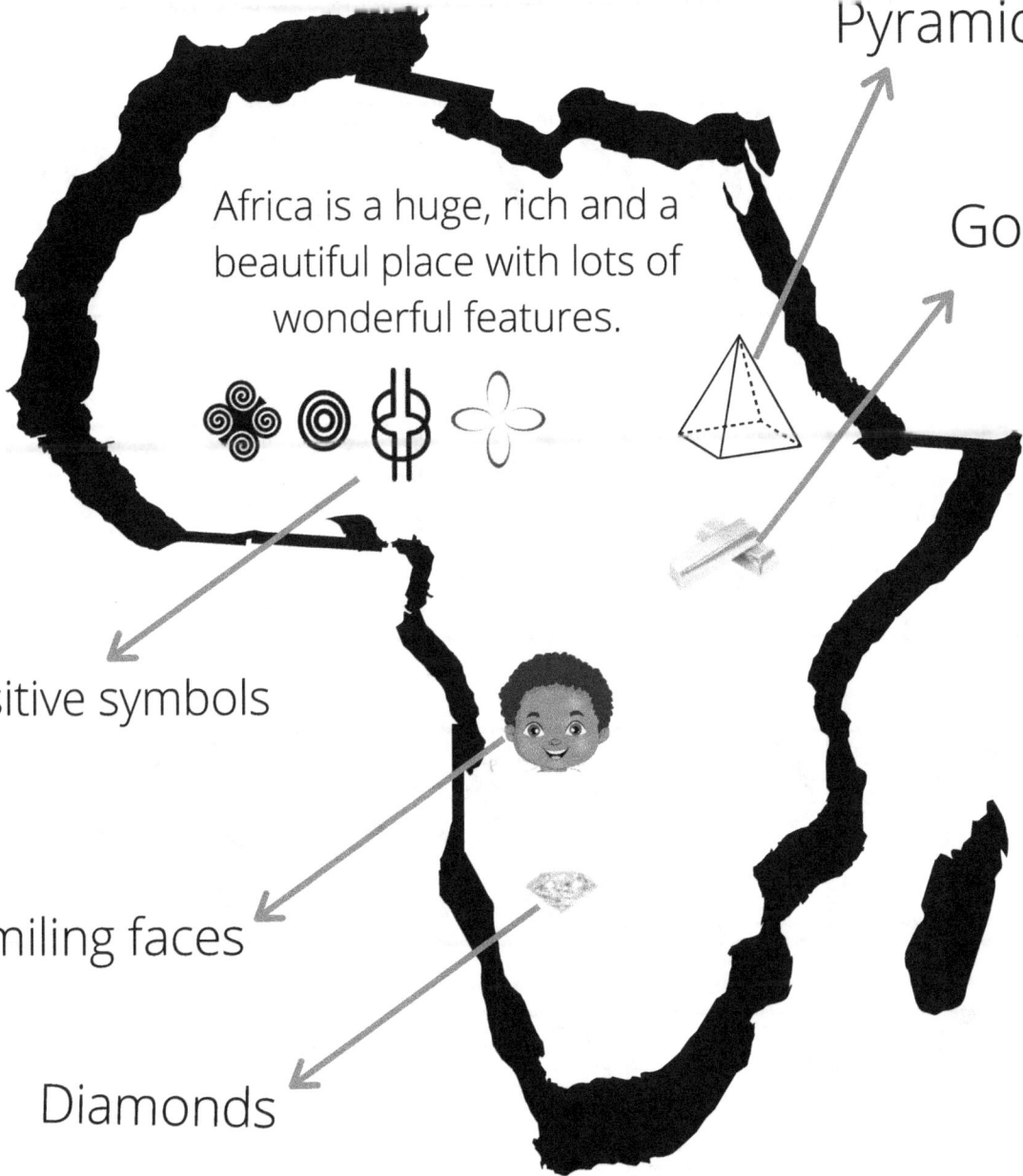

Africa is a huge, rich and a beautiful place with lots of wonderful features.

Pyramids

Gold

Positive symbols

Smiling faces

Diamonds

Counting Numbers 0-20
Practice counting in groups of five (5)

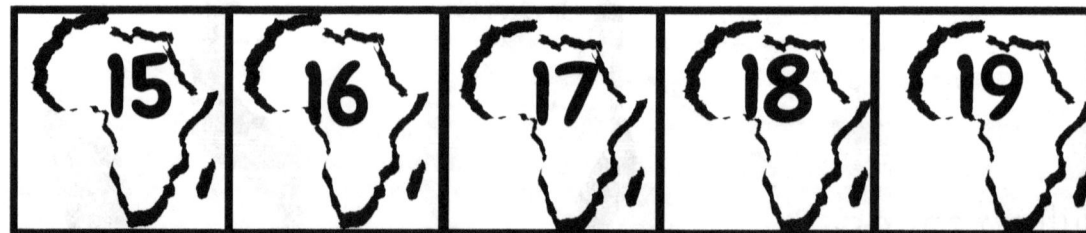

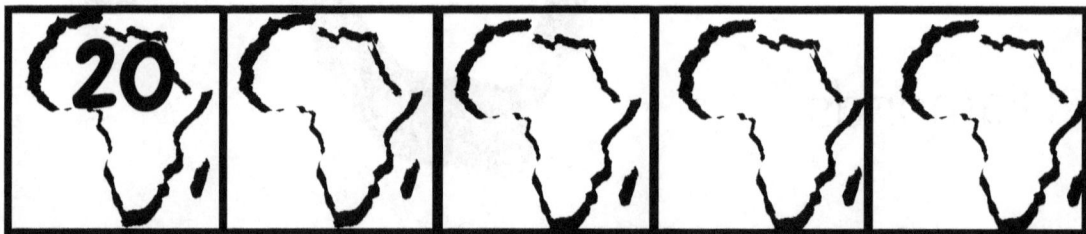

0	1	2	3	4
5	6	7	8	9
10	11	12	13	14
15	16	17	18	19
20				

www.mathjr.org

1. Africa pyramid Egypt gold

2.

1. diamonds smile

2.

Name one of Africa's features

Practice writing numbers 0-4

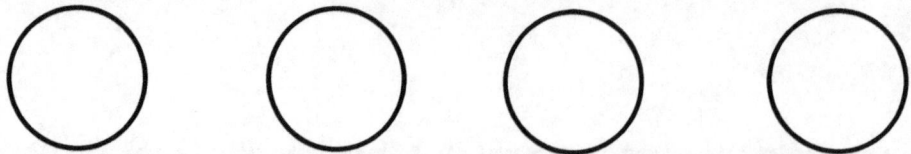

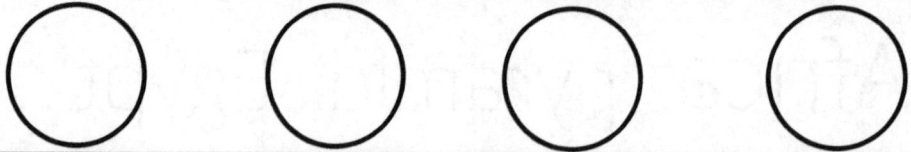

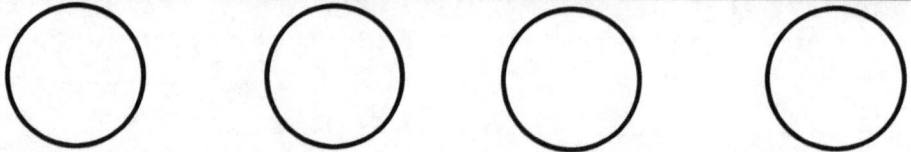

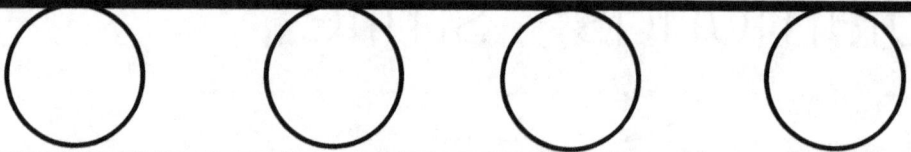

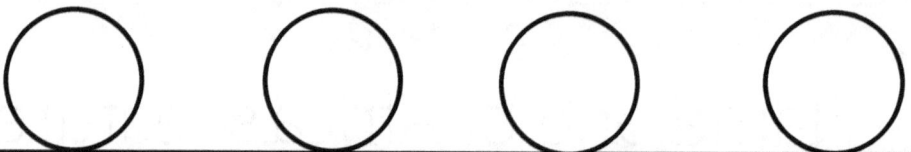

Write the number of African diamonds

  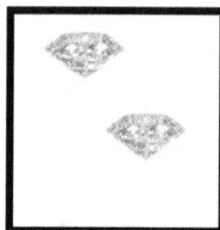 

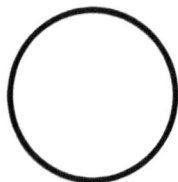 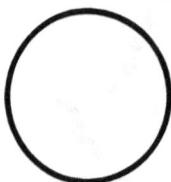  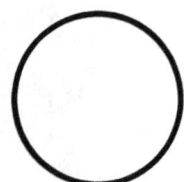

4

www.mathjr.org

Circle the name of the continent below

Asia   Africa   Australia   Antarctica

Circle some of the wonderful features of Africa

diamonds   gold   cold

Pyramids   dolls   smiling faces

5

Pyramids

Pyramids are amazing buildings with the shape of a 4-sided triangle that people of Africa called Egyptians built.

Some are as big as castles.

Egyptians were so smart that no one can figure out how they were able to build such amazing buildings.

See how many pyramids you can find in this book?

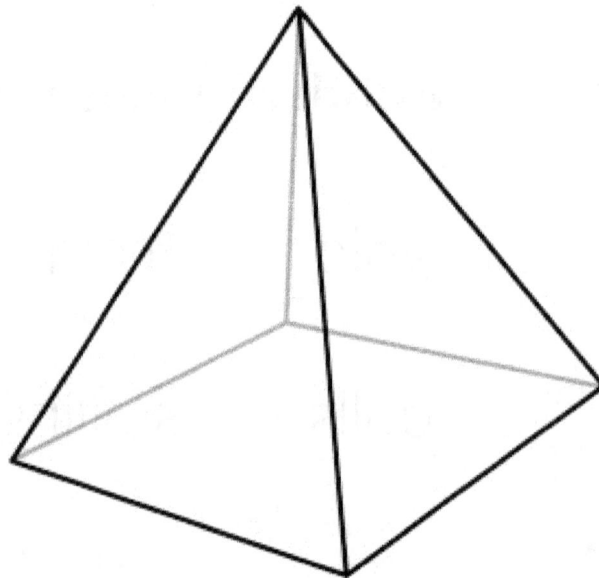

www.mathjr.org

Counting Numbers 0-20

Fill in the missing numbers Count by 5

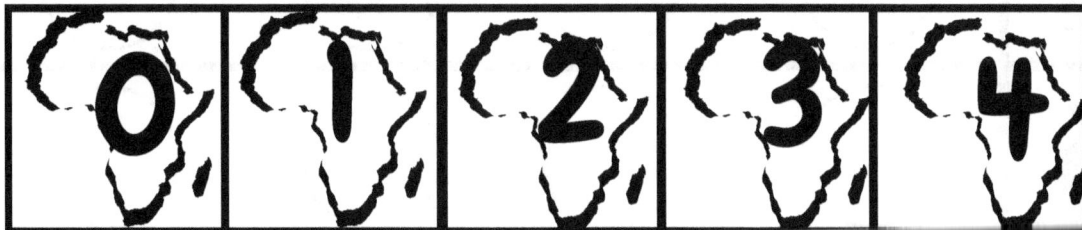

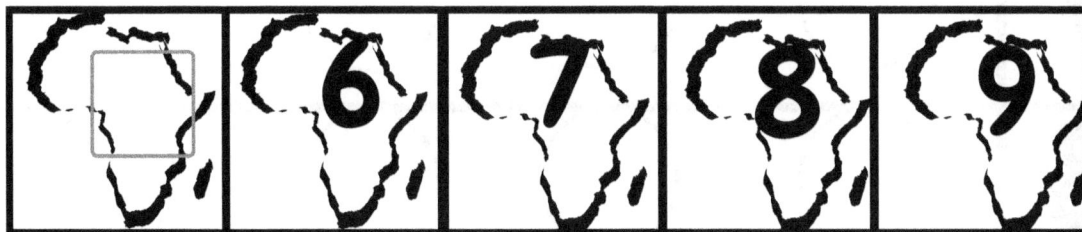

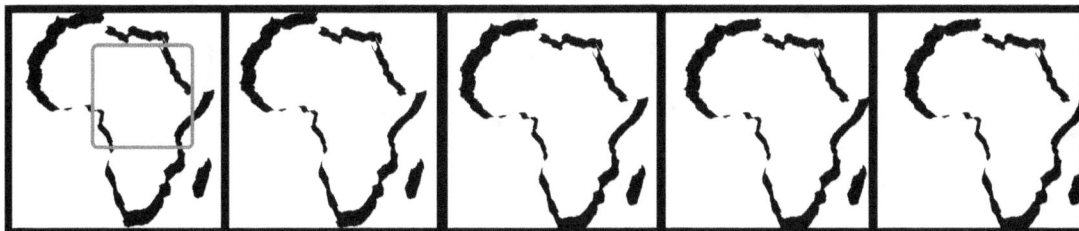

0	1	2	3	4
	6	7	8	9
	11	12	13	14
	16	17	18	19

Practice writing numbers 5-9

5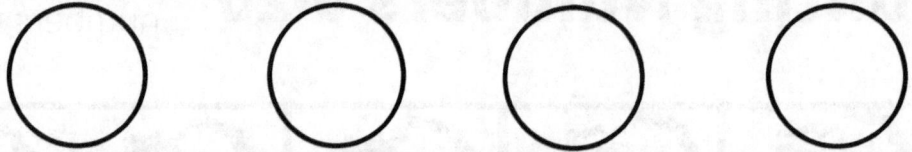

6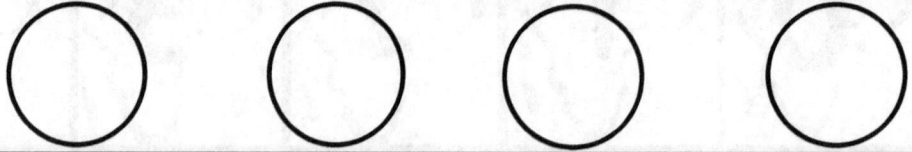

7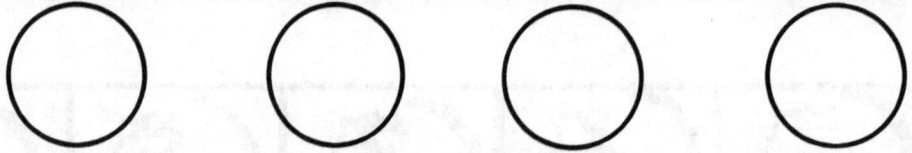

8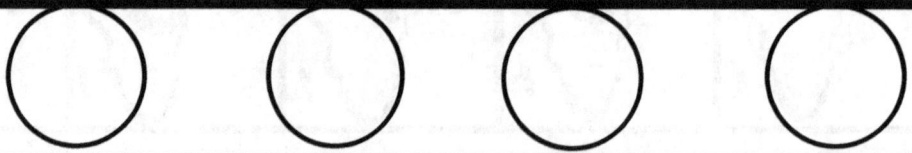

9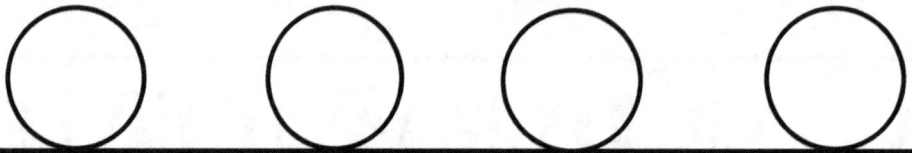

Write the number of African pyramids

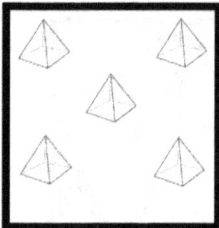 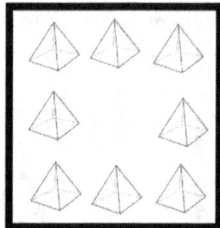 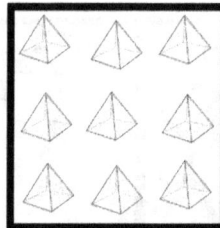 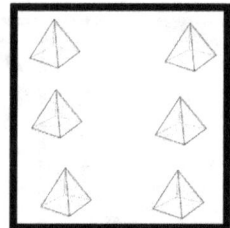

 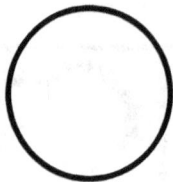 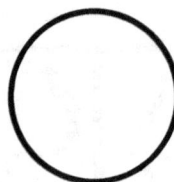 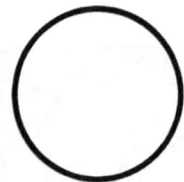

8

Fill in the missing number

0, 1, __, 3, 4, __, 5, 6, 7, 8, __

__, 1, 2, 3, __, 5, 6, __, __, 9

__, __, 2, 3, __, __, _6, __, __, __

0, 1, __, __, __, __, __, __, 8, _

Meet your jr.counter

The jr.counter is a special tool that will help with your math.

Hold your jr.counter up and say: "Hello!"

First lets learn the jr.counter parts!

Upper bead

lower beads

Answer rod

www.mathjr.org

Draw a line to match the part name with the correct part
(and practice writing the word)

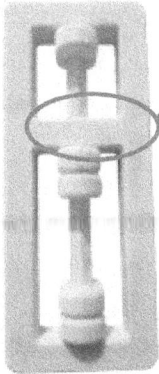

answer rod

1.

2.

upper bead

1.

2.

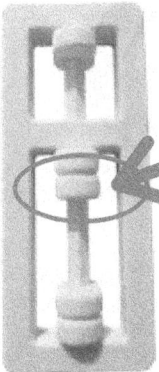

lower beads

1.

2.

Counting Numbers 0-40
Practice counting numbers 0-40

1	2	3	4	5
6	7	8	9	10
11	12	13	14	15
16	17	18	19	20
21	22	23	24	25
26	27	28	29	30
31	32	33	34	35
36	37	38	39	40

www.mathjr.org

Smiling faces

It is so important to be happy and proud of your hard work.

The people who come from African have beautiful smiling faces.

See how many smiling faces you can find in this book.

Color the parts of the jr.counter
1. Color the upper bead **RED**.
2. Color the lower beads **GREEN**.
3. Color the answer rod **BLACK**.

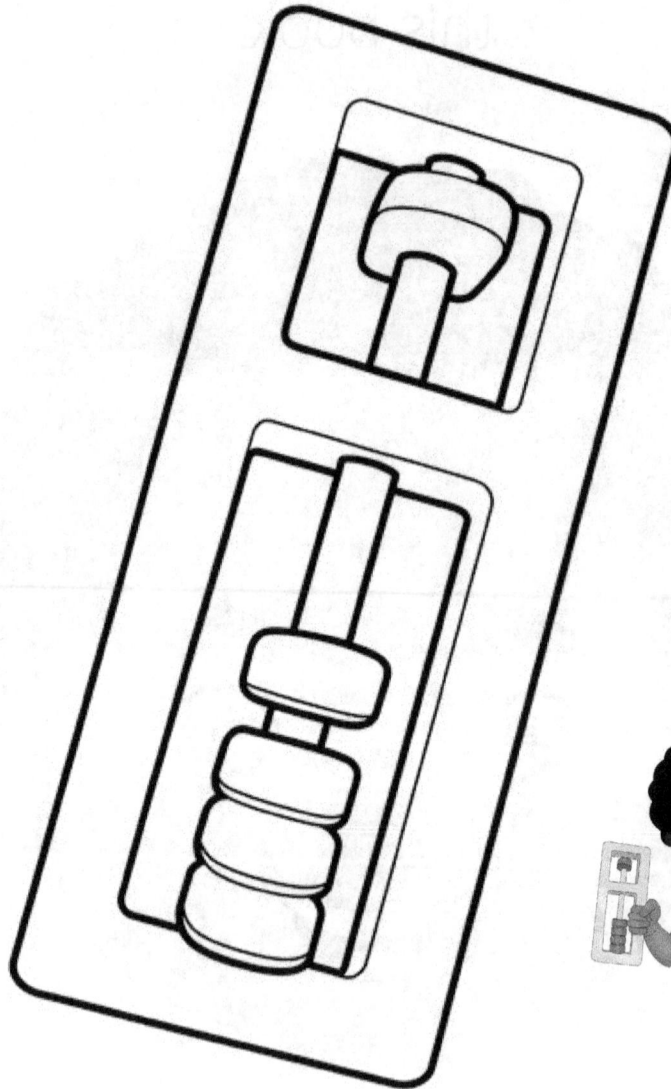

Circle the jr.counter with the matching part

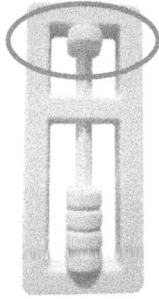 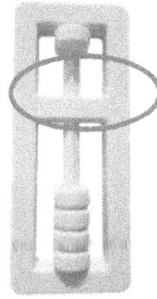 

upper bead

 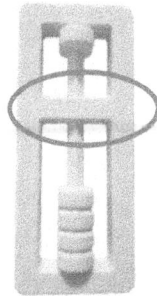 

Answer rod

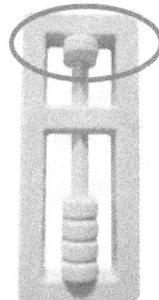 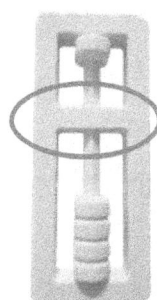 

lower bead

Draw your own jr.counter and circle the lower beads

Draw your own jr.counter and circle the answer rod

Draw your own jr.counter and circle the upper bead

Counting Numbers 0-40

Fill in the missing numbers

1	2		4	
6			9	10
11	12	13		15
16		18		20
	22	23	24	
26			29	30
	32	33	34	
36	37		39	40

Let's learn the positions of numbers 0-4 on your jr.counter!

www.mathjr.org

What do your bead positions mean? How do you position your beads to show numbers 0-4 on your jr.counter?

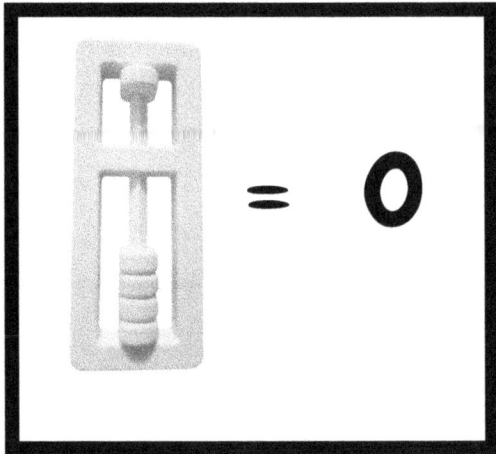 = 0

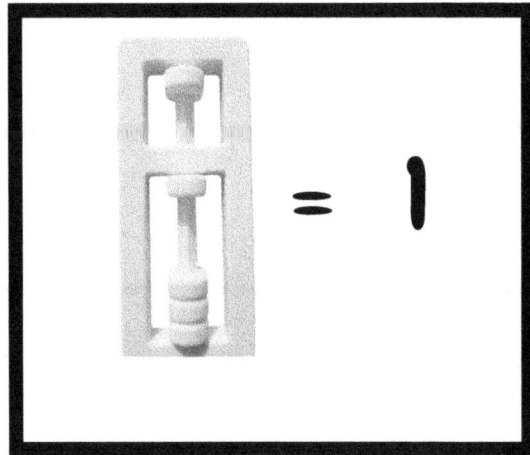 = 1

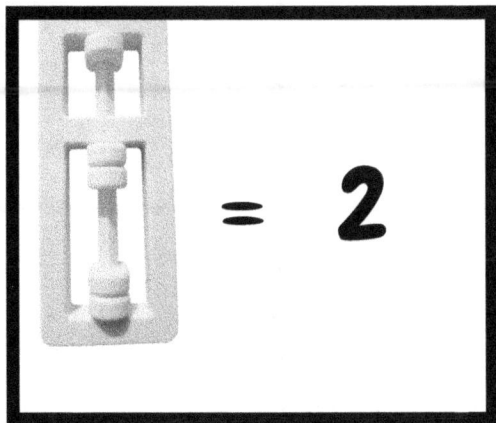 = 2

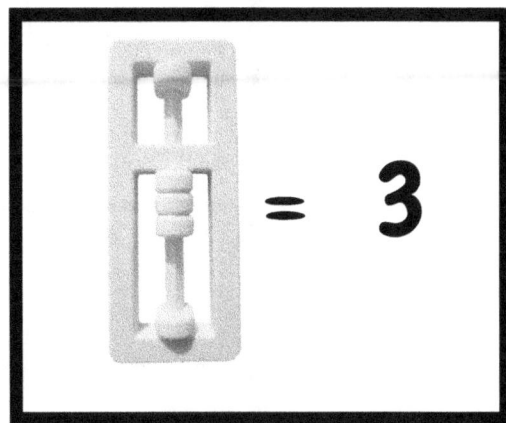 = 3

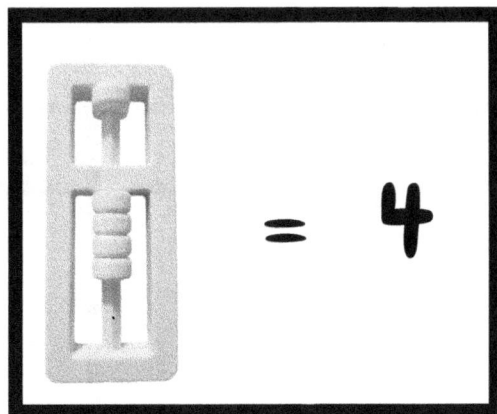 = 4

Practice moving your number positions on your jr.counter.

 = 0
reset

 = 1

 = 2

 = 3

 = 4

 = 0
reset

 = 1

 = 2

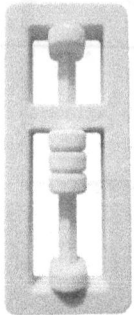 = 3

 = 4

 = 0
reset

 = 1

 = 2

 = 3

 = 4

 = 0
reset

22

www.mathjr.org

Practice moving your number positions on your jr.counter.

= 0 reset	= ___	= 2	= 3
= ___	= 0 reset	= ___	= ___
= ___	= ___	= 0 reset	= 1
= 2	= ___	= 4	= 0 reset

= Equal Numbers =

Equal means the same amount. Look at the different groups below to see the equal values.

Number value	object value	Jr. counter value
3 =		=

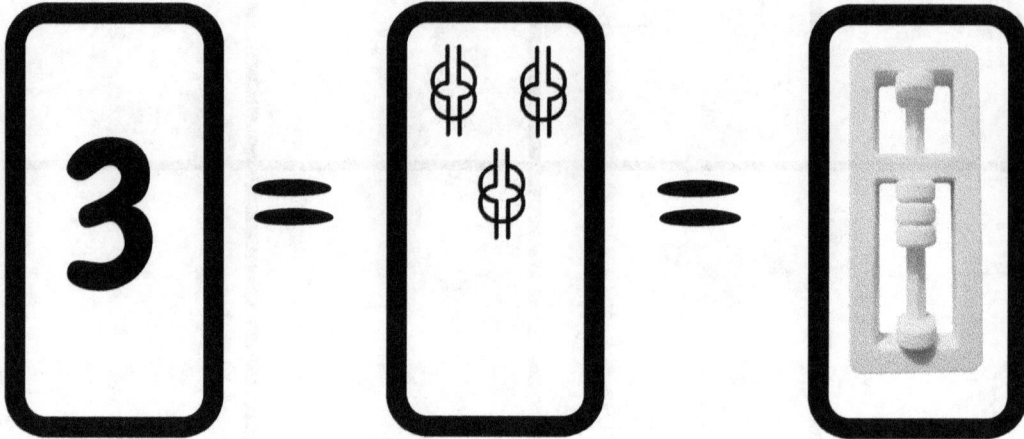

you are very smart

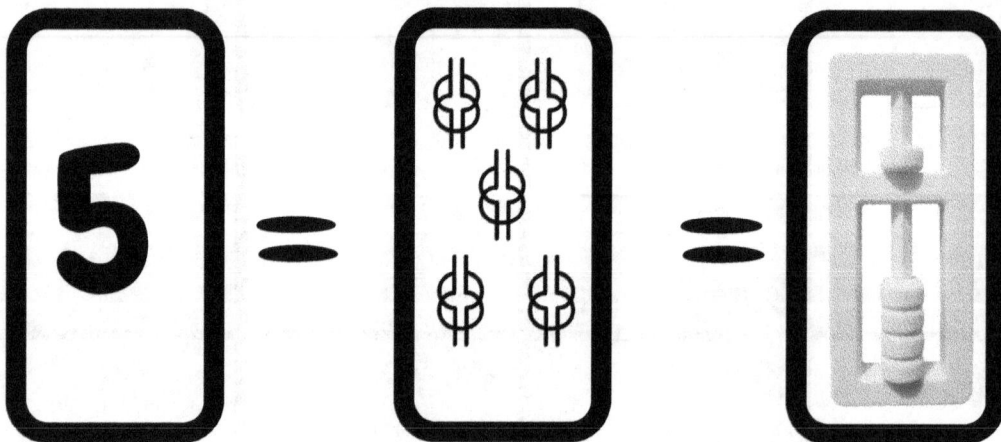

| **5** = | | = |

www.mathjr.org

= Equal Numbers =
Write the number to make an equal number value

= □

= □

= □

= Equal Numbers =
Write the number to make an equal number value

= [ ]

= [ ]

= [ ]

Match the jr.counter Draw a swerved line to match the same number pair

	2
	1
	3
	4
	0

Match the jr.counter
Draw a zig-zag line   WWW   to match the equal number

2	
1	
3	
4	
0	

28

www.mathjr.org

Match the Number pairs
Draw a line to match the same pyramids with the correct jr.counter

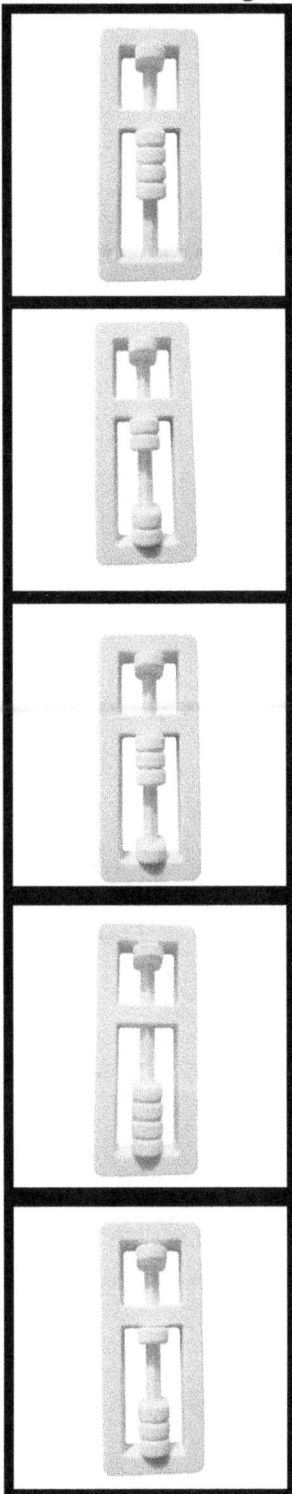

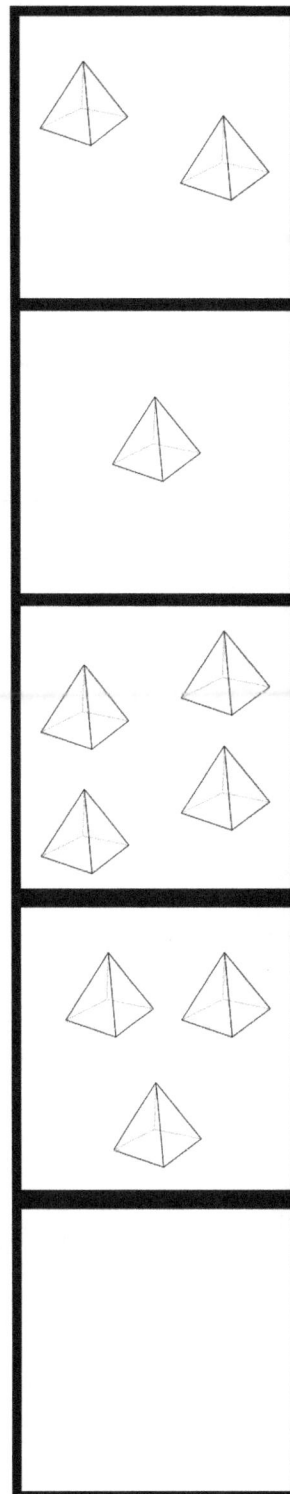

Write the jr.counter position in the circle

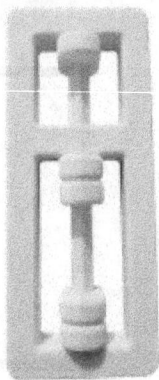

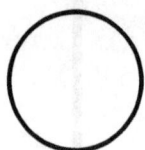

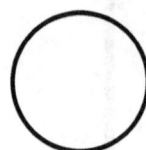

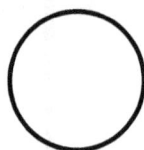

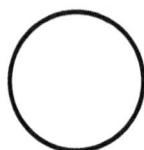

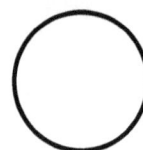

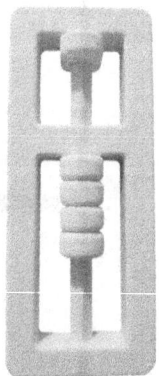

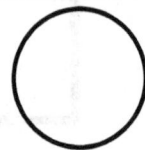

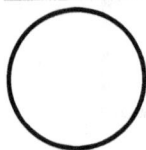

Draw in the correct bead position on the jr.counter

3   0   2   4

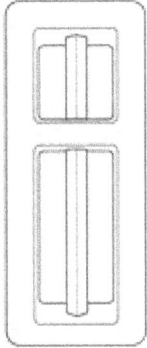 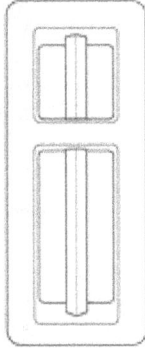 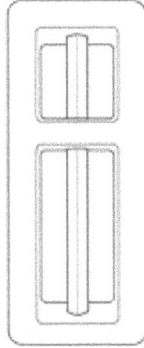 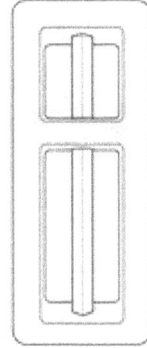

0   1   4   3

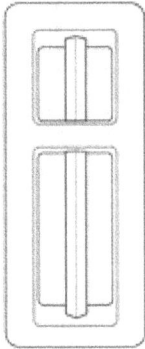 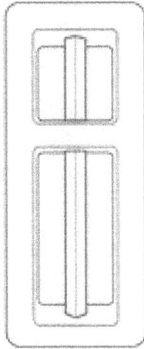 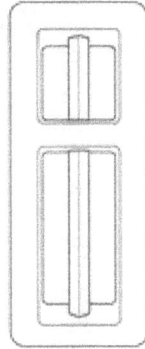 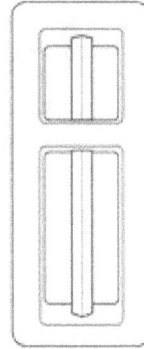

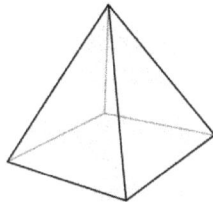

Make your own bead shape
Draw in beads to represent the numbers below

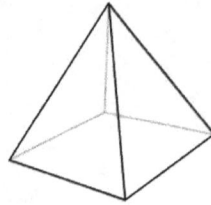

4

3

2

Color your jr.counters

1. Color the jr.counter positioned to the #1 in RED
2. Leave the jr.counters positioned to the #2 BLANK.
3. Color the jr.counter positioned to the #3 in BLUE.
4. Color the jr.counter positioned to the #4 GREEN.

Do you see a green letter ? what letter do you see?

_____

_____

Change the shape of your beads
Draw in beads to represent the number below

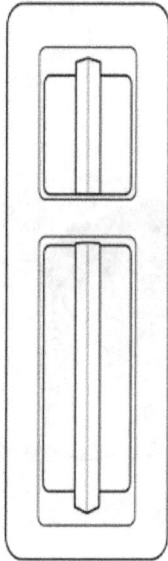

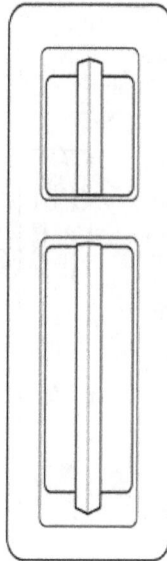

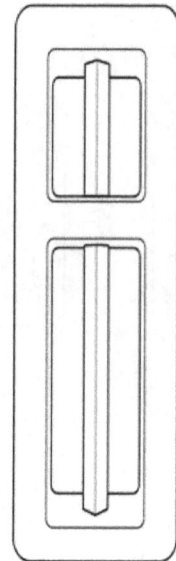

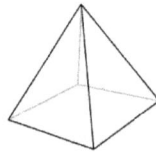

| 1 | 0 | 4 |

Draw your own jr.counter
Draw a HUGE jr.counter positioned in the 0- reset

Draw your own jr.counter
Draw a HUGE jr.counter positioned to number 3

Draw your own jr.counter
Draw a HUGE jr.counter positioned to number 4

Draw your own jr.counter
Draw a HUGE jr.counter
positioned to number 1

What is the number position?
Write the number the jr.counter is showing

Let's learn the number 5 position on your jr.counter!

Number 5

The number 5 bead is the upper bead!

when you move the upper bead down towards the answer rod you jr. counter is at

= 5

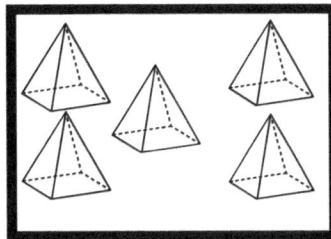

= 5

= 5

= 5

Get to know number 5
Circle the jr.counter positioned to the number 5

 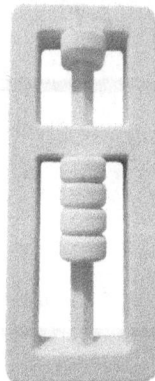  

www.mathjr.org

Get to know number 5
Write the number 5 below the correct counter.

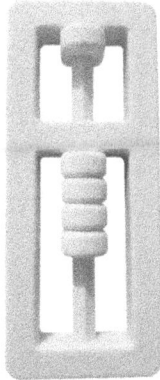

Get to know number 5

1. Color the jr.counter positioned to 5 in RED.
2. Color the jr.counters positioned to 0 reset in BLACK.
3. Color the jr.counters positioned to 1 in GREEN.
4. Leave the rest blank

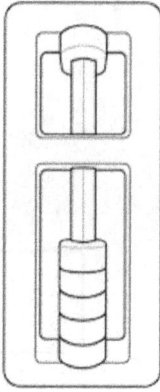

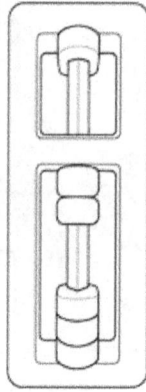

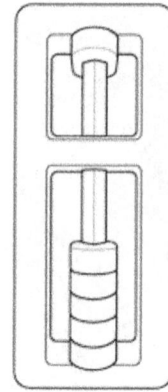

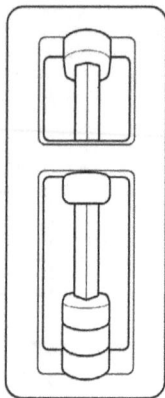

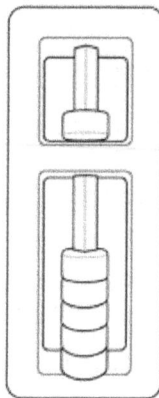

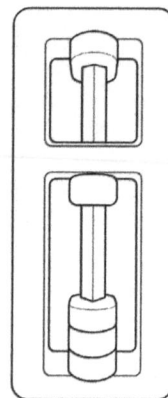

Congratulations!

Your hard work.
Your intelligence.
Your leadership.
Great job completing level 1.
Send us a picture and we will send you your certificate!
ameerah@mathjr.org

Certificate of Completion

9 781735 935409